SOTHIS PRESS

Sothis Press aims to completely reimagine the public domain book with an emphasis on universal access to captivatingly designed print editions. We believe that technology will enable every single book ever published to be perpetually accessible in beautiful print form. Sothis is beauty in ubiquity.

www.sothis.press
1 Waterhouse Square, London, EC1N 2ST

The Florida Hurricane of September 18-20, 1926

Charles Lyman Mitchell

SOTHIS PRESS

ISBN 978-1-0365-0835-7

The Florida Hurricane of September 18-20, 1926 is in the public domain.

This edition is published by Sothis Press The company is registered in England and Wales. The Company Number is 14900389.

UNITED STATES DEPARTMENT OF AGRICULTURE

WEATHER BUREAU

The Florida Hurricane of September 18-20, 1926.

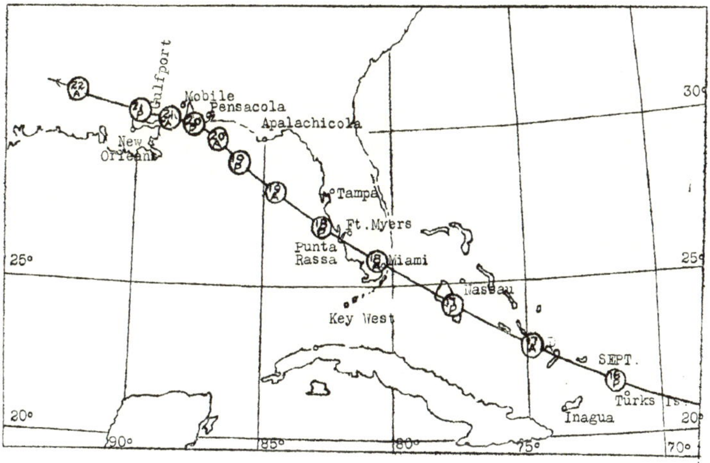

Four days before the center of the hurricane reached Miami (See path and 12-hour movements of center above) it appeared about 200 miles northeast of St. Kitts, W. I. At the same time (8 a.m., Sept. 14) the first tropical disturbance of the month was central about 300 miles southwest of Bermuda and the second was moving slowly northeastward between Cuba and Nassau, Bahamas. At 10 a.m., of the 14th, for the first time in the history of the Weather Bureau, so far as is known, information regarding three tropical disturbances was included in one advisory warning. On the morning of the 15th the following advisory was issued:

"Tropical disturbance reported northeast of St. Kitts Monday morning has moved directly westward. Now centered short distance north St. Thomas, Virgin Islands. This storm has already a[illegible] considerable intensity."

St. Thomas reported a wind velocity of 40 miles an hour from the southwest with a minimum pressure of 29.62 inches, and San Juan reported a wind velocity of 42 miles an hour from the northwest. During the next 24 hours the hurricane moved west-northwestward and the morning of the 16th it was central about 125 miles east-southeast of Turks Island. In the advisory issued that morning it was stated, "The third tropical storm will pass near Turks Island this afternoon or early tonight". A special observation from Turks Island at 11:20 a.m. showed a pressure of 29.62 inches and a wind velocity of 36 miles an hour from the northwest. Another at 1:00 p.m. showed a pressure of 29.26 inches and a wind velocity of 100 miles an hour from the northwest. This was the last report received from Turks Island. Undoubtedly much damage was done on this island. Sending of regular observations was not resumed until October 6. At 3 p.m. of the 16th the following advisory warning was issued and given wide distribution:

"Center of hurricane of great intensity passing near Turks Island which reports wind one hundred miles from northwest. Hurricane center will pass near or slightly north of Crooked Island, Bahama group, Friday forenoon. Greatest caution advised vessels bound for Bahama group and adjacent waters."

At 9:30 p.m., the following advisory was issued:

"* * * Third tropical storm has passed Turks Island moving west-northwestward attended by dangerous shifting gales. Caution advised vessels bound for Florida Straits, Bahamas and adjacent waters."

At 10:20 a.m. of the 17th the following warning was issued:

"Hoist northeast storm warning twelve noon Jupiter Inlet to Key West. Hurricane central about twenty-three north seventy-four west moving west-northwestward attended by winds hurricane force near center. This is very severe storm. Its center will likely pass near Nassau early tonight. Great caution advised all vessels bound Florida Straits, Bahama Islands and east Florida coast. Every precaution should be taken for destructive winds Saturday morning especially Jupiter to Miami."

To the Governor General, Nassau, Bahamas, the following message was sent:

"Please send special observations every two hours today. Hurricane central near and north Crooked Island and its center will likely pass near Nassau early tonight. This is a destructive storm."

Only two special observations were received from Nassau. The first was taken at 10 a.m., showing a pressure of 29.64 inches and a wind velocity of 34 miles an hour from the northeast, and the second, at 1 p.m., when the pressure was 29.56 inches and the wind was 50 miles an hour from the north. The lowest barometer reading and the maximum wind velocity at Nassau, as well as the amount of damage done at Nassau are unknown at this writing. However, press dispatches from that place since the hurricane indicate that the storm was about as severe as the destructive hurricane of July 25-26, this year, but that it did not last as long. One dispatch stated that the wireless towers which were seriously damaged in the July hurricane were partially dismantled before the present storm became very severe, which probably accounts for the fact that no special observations were received after 1 p.m.

Because, no doubt, of the heed given the advisory warnings by vessel masters, no vessels ventured close to this hurricane. Consequently, no vessel reports of material value were received at any time during the progress of the hurricane from longitude 60°W until it moved inland west of Mobile, Ala. After the last special observation from Nassau at 1 p.m., of the 17th, no reports whatever were received from the region east of Miami and Key West. At 8 p.m. both of these stations reported a barometer reading of 29.68 inches, with northeast wind, 18 miles an hour at Miami and 12 at Key West. Furthermore, both stations reported a pressure increase of 0.04 inch within the last two hours. The lack of any information from the region to the eastward and the rather disconcerting reports of a pressure rise at these two stations placed the forecaster in a very difficult position. With night already on and no chance of awaiting special reports from Miami, he had to rely on his previous calculations, made Friday morning, which placed the hurricane center near Miami at 8 a.m., Saturday morning. Therefore, with no indications whatever of a recurve in the path of the hurricane, the storm warnings were changed to hurricane warnings at 11 p.m. of the 17th from Jupiter Inlet to Key West, and northeast storm warnings were ordered north of Jupiter to Titusville and north of Key West to Punta Gorda. As there had been plenty of time during the daylight hours for making all possible preparation for the hurricane in the Jupiter-Miami section after the receipt of the significantly worded warning of Friday morning, there was not much left that could be done in that section after the actual display of hurricane warnings.

The 8 a.m. observation of the 18th at Key West showed a pressure of 29.52 inches and a wind velocity of 38 miles an hour from the northwest, and the Ft. Myers report showed a pressure of 29.46 inches and a wind velocity of 30 miles an hour from the northeast. These reports showed clearly that the center of the hurricane was very close to Miami and still moving west-northwestward. Hurricane warnings were ordered displayed at 10 a.m., north of Key West to Tampa and south of Titusville to Jupiter, and northeast storm warnings north of Titusville to Jacksonville and north of Tampa to Mobile, Ala.

At 1:30 p.m., the following warning was issued:

"Hoist northeast storm warning 4 p.m. north of Jacksonville to Charleston and west of Mobile to mouth of Mississippi River. Hurricane center noon over extreme southern Florida., Fort Myers reporting barometer 29.04 wind fifty-two miles north. Hurricane will pass into Gulf of Mexico this afternoon and continue to move west-northwestward for the present. This is a very severe storm. Greatest caution advised vessels in its path."

At 9:45 p.m., of the 18th, the following warning was issued:

"Change to hurricane warning 11 p.m. Apalachicola, Fla., to Burrwood, La. Hurricane central between twenty-six and twenty-seven north and about eighty-three west moving west-northwestward attended by winds of hurricane force. This is a very severe storm. Unless course changes hurricane center will move inland most likely between Pensacola and mouth of Mississippi River Sunday night. Emergency. Every precaution should be taken against destructive winds."

The next morning the following was issued:

> "Advisory 10 a.m. * * * Hurricane apparently central between twenty-seven and twenty-eight north and about eight-five west moving west-northwestward attended by dangerous shifting gales. Unless course changes hurricane center will move inland late tonight between Pensacola and mouth of Mississippi River, probably nearer the latter. Further advices this afternoon. Meanwhile every precaution should be taken against destructive east and northeast winds beginning tonight all points where hurricane warnings are displayed.

At 2:30 p.m., of the 19th, the following advisory warning was sent to all stations from Apalachicola to Burrwood, inclusive, and hurricane warnings were ordered continued at 11 p.m. at all display stations within this area:

> "Noon specials indicate hurricane center near twenty-eight north eighty-six west moving west-northwestward. This is a hurricane of great intensity and magnitude and emphasis should be placed on need of every possible preparation for destructive winds, especially Pensacola to mouth of Mississippi River. Hurricane center will likely pass inland late tonight or Monday morning."

Then the final advisory before the hurricane center approached the coast was as follows:

> "Advisory 9:30 p.m. Hurricane central about twenty-nine north eighty-seven west apparently moving northwestward. Hurricane center will pass inland early Monday morning probably not far from the Pensacola-Mobile section."

Instead of passing inland in the Mobile-Pensacola section, the hurricane center, after advancing northwestward with diminishing rate of movement to a position a short distance southwest of Pensacola, turned toward the west and moved very slowly westward south of Mobile and along the Mississippi coast. Its center passed Gulfport, Miss., about 9 a.m. of the 21st, but the storm had diminished greatly in intensity by this time. At 8 p.m., the center was some distance northwest of New Orleans. During the ensuing twenty-four hours the disturbance moved more rapidly westward over Louisiana into eastern Texas where it dissipated.

This hurricane, although it was first noted northeast of St. Kitts, W. I., on the 14th, doubtless originated far to the eastward, most likely to the southwest of the Cape Verde Islands, some six or seven days earlier. Nearly all of the severe tropical cyclones of the North Atlantic Ocean during August and September originate far to the east of the Lesser Antilles in the belt of doldrums where this belt is farthest north, not far from the Cape Verde Islands. (See pp. 16-17, Sup. #24, Monthly Weather Review). In the first three days following its appearance northeast of St. Kitts, the hurricane center moved about 1000 miles, or at a rate of about 14 miles an hour; during the 24 hours before its arrival at Miami, about 450 miles, or 18.75 miles per hour, which is unusually rapid; and during the next two days when it was crossing southern Florida and the northeastern Gulf of Mexico, about 500 miles, or only 10.4 miles an hour. Between 8 a.m. of the 20th and 8 a.m. of the 21st, the center advanced only about 150 miles, or 6.25 miles an hour.

This is one of the most severe hurricanes that ever reached the coast of the United States. The damage to property was undoubtedly far greater than in any other hurricane, while the loss of life has been exceeded three times, as follows: Galveston, Tex., September 8, 1900, about 6,000; Georgia and South Carolina coasts, August 28, 1893, about 2,000; and on the Louisiana coast, October 2, 1893, about 1,500. No accurate estimate is yet possible of the number of lives lost in the recent hurricane in southern Florida, but, according to a rough estimate of the official in charge at Jacksonville, Fla., it may reach or exceed 1,000. The known loss of life in the Miami district is 114, and at Moore Haven, 200. It is remarkable that there was no loss of live either in Pensacola or Mobile, although the wind at Pensacola attained a maximum velocity of 120 miles an hour for a period of 10 minutes, and the fastest mile was at the rate of 152 miles an hour, and was above 100 miles most of the time for four hours, while it reached 94 miles an hour at Mobile.

This is the sixth tropical cyclone in 40 years to pass inland over the southeastern Florida coast and cross the peninsula into the Gulf of Mexico, and, of course, this was by far the most severe of the six.

It is quite noteworthy that the center of the hurricane passed over or near several cities or towns where accurate pressure readings were taken. The center reached the southeast Florida coast at Miami about 6:45 a.m., September 18, and there was a lull in the wind of about 35 minutes. Estimating the rate of progression at this time as 18 miles an hour, the diameter of the center or "eye" of the storm was about 10-1/2 miles. The lowest pressure was 27.61 inches, this being the lowest ever registered in the United States. Nearing the west coast of extreme southern Florida, the center passed over Bonita Springs, about 20 miles south of Fort Myers, shortly after noon. It passed into the Gulf of Mexico during the afternoon, the displayman at Punta Rassa reporting a calm at 3:15 p.m., and the lowest pressure, 28.05 inches, at 3:30 p.m. After crossing the northeastern portion of the Gulf, the center closely approached, but did not reach, Pensacola at 3:10 p.m., September 20, when the pressure was 28.56 inches. Passing toward the west a short distance south of Mobile at 9:30 p.m., when the barometer read 28.76 inches, the center passed a very short distance south of Pascagoula, Miss., at 5:25 a.m., of the 21st when the pressure was 28.99 inches. The northern edge of the "eye" of the storm passed over Biloxi about 8 a.m., with a pressure of 29.03 inches at that time at 9 a.m., the center reached Gulfport and there was a calm of about 10 minutes during which the barometer read 29.08 inches. The center passed over Pass Christian about 9:30 a.m., and there was a calm of about 30 minutes. The lowest pressure was about the same as at Gulfport. Beginning about 9:30 a.m., there was calm for an hour at Bay St. Louis. The hurricane center moved over land after leaving Pass Christian and Bay St. Louis.

These barometer readings show that the hurricane gradually diminished in intensity after passing inland at Miami, but it was still of great intensity until after it passed westward south of Mobile.

Below are reproduced copies of the barograph traces (corrected to sea-level) made at Miami and Mobile during the hurricane. The marked difference in the appearance of the two traces is due to the fact that the hurricane was moving quite fast while approaching Miami and advancing west-northwestward from that place, while it was moving quite slowly, as well as diminishing in intensity, while approaching Mobile and recurving to the westward south of that city.

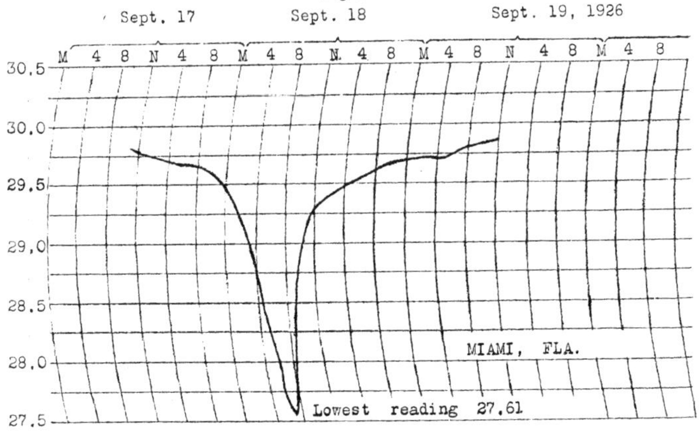
Sept. 17
Sept. 18
Sept. 19, 1926
M 4 8 N 4 8 M 4 8 N 4 8 M 4 8 N 4 8 M 4 8
30.5
30.0
29.5
29.0
28.5
28.0
27.5
MIAMI, FLA.
Lowest reading 27.61

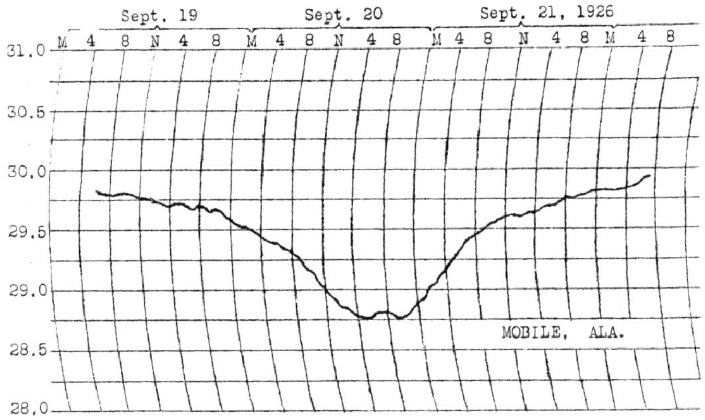
Sept. 19
Sept. 20
Sept. 21, 1926
M 4 8 N 4 8 M 4 8 N 4 8 M 4 8 N 4 8 M 4 8
31.0
30.5
30.0
29.5
29.0
28.5
28.0
MOBILE, ALA.

The following are parts of the reports on the hurricane rendered by the officials in charge of the stations affected by the storm:

Miami, Fla. The first information concerning the storm was received from the Central Office at 11:30 a.m. of the 14th. No vessels bound for the Bahamas left Miami after that date. Advisory messages relative to the intensity and progress of the storm were received at regular intervals from the 15th to the 17th, inclusive, and these advices were given such wide distribution that it can be safely said that the entire population of the lower east coast of Florida was informed of the approach of the storm.

Northeast storm warnings were displayed, by order of the Central Office, at noon of the 17th. The afternoon newspapers published the warning, and it was otherwise disseminated by telephone and telegraph. From the early afternoon of the 17th until the wires were blown down, telephone calls at the Weather Bureau office were answered at the rate of two to three per minute. In addition to the telephone service from the Weather Bureau, the Miami Daily News kept a special telephone operator on duty to give information to those who did not succeed in getting telephone connection with the Weather Bureau. A representative of the News remained at the Weather Bureau office throughout the night of the 17th-18th and kept his paper informed of all available information until telephone connection was severed.

The message ordering hurricane warnings at 11 p.m. of the 17th was received at 11:16 p.m. The warning was displayed from the roof of the Federal Building at 11:25 p.m., and from the storm warning tower at the city docks, one and one-half miles from the Weather Bureau office, at midnight. Before leaving for the storm-warning tower, I gave the hurricane warning to the long distance telephone operator, who repeated it to the telephone exchanges at Homestead, Dania, Hollywood, and Fort Lauderdale. The warning was also telephoned to the chief dispatcher of the Florida East Coast Railroad, and several efforts were made to get telephone connection with Fowey Rock Lighthouse and the Coast Guard base at Fort Lauderdale. Telephone communication had not been interrupted, but the operator reported that repeated calls failed to get any response from Fowey Rock or the Coast Guard station. Shortly after 10 p.m. I began to give out the information that the rapid fall of the barometer and the direction and increasing velocity of the wind indicated that the storm was rapidly approaching this coast, and that, unless it recurved to the east of Miami, winds of hurricane force might be expected. This information continued to be given by telephone until the receipt of the hurricane warnings at 11:16 p.m. After that time all persons calling by telephone or in person were informed of the display of hurricane warnings. Telephone communication with Hollywood and Miami Beach was severed between 1 a.m. and 2 a.m., and in Miami between 2 a. m. and 3 a.m.

The hurricane came with great suddenness. Except for a moderate but steady fall of the barometer after 10 a.m. of the 17th, there were no unusual meteorological conditions to herald the approach of the storm. The wind velocity as late as 8 p.m. of the 17th was only 19 miles per hour, and the usual heavy rain that precedes a tropical storm did not set in until after midnight, by which time the wind was blowing a fresh gale. At 10 p.m. of the 17th the barometer began to fall rapidly, and by midnight it had fallen 0.11 inch. From midnight to 6:45 a.m., at which time the center of the center of the storm passed over Miami, there was a precipitate fall at the rate of 0.28 inch per hour. *****From about 5:30 to 6:10 a.m. the barometer fell 0.40 inch and then remained stationary for 15 or 20 minutes. This was at the beginning of the lull in the wind that attended the arrival of the center of the storm. After the short stationary period there was another rapid fall of 0.06 inch, and at 6:45 a.m., a reading of the mercurial barometer showed a pressure of 27.61 inches. ****After the passage of the center of the storm, the barometer rose even more rapidly than it had fallen, and by noon it had reached 29.30 inches.

The center of the storm passed over the central and southern parts of Miami. Over the extreme northern part of the city and over the northern part of Miami Beach the wind shifted from northeast to south, but there was no pronounced lull. At the Weather Bureau office the wind fell to 10 miles per hour at 6:30 a.m. At the same time the velocity at the Allison hospital in the northern part of Miami Beach, was 80 miles per hour. Ten minutes before, the velocity had been 108 miles.

It will be necessary to describe the exposure of the anemometer at the Weather Bureau office in order that the low wind velocities recorded may be understood. The anemometer is located on the 3-story Federal Building, and is almost completely surrounded by buildings of 8 to 18 stories. One 15-story building is only 100 feet east-northeast of the anemometer. Another 17-story building is about 250 feet due east. The average velocity of northeast winds prior to the erection of these buildings was 1.44 times the average velocity of northeast winds from January to August, 1926. Since the completion of the walls of the 15-story buiding, within the last two months, there has been a still further reduction in the velocity of northeast winds. The multiple 1.4, however, has been used in correcting the velocities recorded during the first phase of the storm, when the wind was northeast.

The wind increased steadily from the northeast after 10 p.m. At 1:50 a.m the anemometer recorded a velocity of 41 miles, indicating a true velocity of about 57 miles per hour. Telephone communication with Miami Beach ceased shortly before this time. By 2:35 a.m. the true velocity had increased to 60 miles per hour, and by 3 a.m. telephone service in Miami had ended. There was a steady increase in wind velocity from that time to 5 a.m. when the anemometer recorded a maximum velocity of 80 miles, indicating a true velocity of at least 115 miles per hour. The top of the raingage blew off at 3:42 a.m., and was recovered and replaced by the assistant at this station. It was again blown off a few minutes later and lost. A part of it was found the next day on the roof of a nearby building. The electric light wires were blown down at 4 a.m., and the observations during the remainder of the night were made with a flashlight, supplied by one of the visitors that spend the night at the Weather Bureau office. Frequent flashes of electricity from fallen wires added to the fearful aspect of the elements. The instrument shelter blew away between 4 a.m. and 5 a.m., landing in the street below and crashing into the automobile of a Miami Daily News staff writer who was on duty at the Weather Bureau office. There was an abrupt decrease in the wind velocity between 6:10 a.m. and 6:15 a.m., when the center of the storm reached Miami. Many persons who had spent the night in down-town buildings rushed out to view the wreckage that filled the streets. I warned those in the vicinity of the Federal Building that the storm was not over and that it would be dangerous to remain in the open. The lull lasted 35 minutes, and during that time the streets of the city became crowded with people. As a result, many lives were lost during the second phase of the storm. With the passage of the center of the storm, the wind shifted to southeast at 6:47 a.m., and immediately increased to gale force. A velocity of 50 miles was recorded at 7:55 a.m. and a velocity of 60 miles at 8:55 a.m. These recorded velocities are nearly 50 per cent less than the actual velocities. The wind shifted to southwest at 9 a.m. and continued from that direction until 6 p.m., with steadily diminishing force.

A Robinson anemometer on the roof of the Allison Hospital, Miami Beach, connected with a Weather Bureau type triple register, recorded a velocity of 128 miles per hour at 7:30 a.m. The anemometer blew away at 8:12 a.m., at which time it was recording 120 miles per hour.

The storm tide on the Miami side of Biscayne Bay was approximately eight feet, and reports indicate a similar tide at Miami Beach. The water front of Miami was flooded for two to three blocks back from the bay, and low parts of the city near the Miami River were also flooded. After the storm, the entire bay frontsection of Miami was strewn with boats ranging in size from small pleasure craft to large schooners. Some of the boats had been carried more than two blocks from the bay. Water rose in hotels and residences near the bay to a depth of three to five feet. Miami Beach was entirely inundated, and, at the height of the tide, the ocean extended to Miami, three and one-half miles across Biscayne Bay. All streets near the ocean at Miami Beach were covered with sand to a depth of several feet, and in some places automobiles were entirely covered. The foundations of some buildings were washed out, allowing the buildings to collapse. The storm tide occurred with the shift of the wind to the east and southeast, following the arrival of the center of the storm. In the Miami River, the tide came in the form of a bore that left a mass of wreckage from the boats that had sought safe anchorage.

The intensity of the storm and the wreckage that it left cannot be adequately described. The continuous roar of the wind; the crash of falling buildings, flying debris, and plate glass; the shriek of fire apparatus and ambulances that rendered assistance until the streets became impassible; the terrifically driven rain that came in sheets as dense as fog; the electric flashes from live wires have left the memory of a fearful night in the minds of the many thousands that were in the storm area.

The known loss of life in the Miami district is 114. Many more are missing. Several thousand persons were injured, and 25,000 were without shelter after the storm.

The property loss in the greater Miami area has been estimated at $76,000,000. This does not include damage to house, office, and store furnishings. Approximately 4725 homes were destroyed and 9100 damaged in the area extending from Fort Lauderdale to Miami.

On September 28 the marine unit of the Miami citizens' committee estimated that 402 craft were sunk or wrecked in the Miami area during the hurricane. A survey of Biscayne Bay and the Miami River revealed 202 boats sunk or wrecked, exclusive of 200 other boats lying high and dry along the bay front and in Royal Palm Park. In the waters of the bay lie the remains of 58 craft, including 3 sloops, 20 barges, 20 yachts, 2 motor boats, a dredge, 3 houseboats, 1 tug, 1 rigged yawl, a converted subchaser, and 6 schooners, 2 of which were four-masters. Most of the 144 craft sunk or wrecked in the river were of light draft, including 51 houseboats, 22 yachts, 46 motor boats and launches, 10 barges, 6 scows, 3 lighters, a fishing smack, 2 derricks, a tug, a dredge, and 1 oil boat. Almost every houseboat in the river was believed occupied and their loss rendered many homeless.

On the same date (September 28) the American Red Cross officials estimated the number of homeless families in the storm-stricken area of southern Florida at 15,700. With a conservative average of three persons to a family, the number of needy refugees was believed to be more than 47,100. Many hundreds of other families, rendered without shelter, left the scene without reporting their losses.

Jacksonville, Fla.- **********Effective on the 14th, all advisory and storm warning orders were given prompt distribution through the medium of the regular storm warning and small craft warning stations; the Naval Radio stations at Jupiter and St.Augustine; the local Municipal Radio station, WJAX, and through the cooperation of the telegraph departments of the A.C.L., the S.A.L., and Florida East Coast railways.

******A message was received from the Naval Radio station, Jupiter, at 6:33 p.m. on the 17th to the effect that the Nassau station failed about noon; no answer to calls. Previously, however, all messages reached Nassau, hence timely information regarding the location and progress of the hurricane was in the hands of the proper authorities.

******It appears that the heaviest rainfall occurred on the Gulf side of Florida, rather than along the Atlantic seaboard, the greatest amount 16.40 inches, occurring at Blountsown during the 20th-21st. The lowest pressure at Jacksonville was 29.77 inches, occurring at 4 a.m. on the 19th. The highest wind velocity was 50 miles from the east at 2:10 p.m. of the 18th. There was a verifying velocity from 9:52 a.m. to 7:50 p.m. of the 18th.

******Considering individual units, Moore Haven, Glades county, suffered in mortality, relatively, more than any other town or city, as the result of the breaking of the dikes on the south-southwest side of Lake Okeechobee. The height of the waters in the lake depends upon wind direction and velocity. The dikes were supposed to be secure in every respect, but when the north-northeast hurricane winds lashed the lake the storm wave swept over the dikes tearing them asunder, and fell upon the helpless population, then possibly as much as six feet below the storm wave, which in comparatively low places, meant the submerging of women and children.

It is assuring to know that, as early as the 16th, this office telegraphed to the local engineer of the State Drainage Board, and, who is, also, the storm warning displayman and observer at the second order station at

Moore Haven, that north-northeast winds would continue on the lake, probably increasing, the message containing in addition the location and movement of the tropical storm as indicated by the Central Office; and at noon on the 17th northeast storm warnings were ordered for Moore Haven and Canal Point. Loci on the east side of Lake Okeechobee did not feel the impact of the storm wave as did those on the south-southwest side. ******

The greatest damage to crops occurred from about St.Lucie county on the east coast westward to the Gulf coast line; again from Escambia eastward, when the storm struck the west coast of the State after passing over the lower portion of the peninsula. The damage to the several crops in the order of their importance is:

Citrus fruits in the peninsula (chiefly grape fruit) $10,000,000. Truck, corn, cotton, cane,peanuts, sweet potatoes greatly damaged, especially near storm track (No estimate of damage) Large areas timber blown down, and many saw mills prostrated(No estimate) Shipping, all classes, large and small chiefly on the lower coast, thence westward....................5,000,000.

Tampa, Fla. - *****On the morning of the 18th northeast storm warnings were ordered from Punta Gorda to Cedar Keys, and warnings were changed to hurricane north of Key West to Tampa a short time later. After the flags were hoisted, the emergency telephone list was called, and telephone, electric, and gas companies were notified, also the Mayor, the Chief of Police, the Fire Department, and the Florida Airways. Long distance calls were sent to the beaches. Another phone had to be used in sending out warnings, as immediately upon hoist of the flags incoming calls clogged the line, and did not cease until the night of the 19th. The forecast card (in red), printed about 11:30 a.m., carried the warning. Special bulletin cards in red printed at 2:30 p.m. and 6:30 p.m. gave further particulars. It was stated that winds locally would be between 50 and 60 miles per hour, but that there would be no dangerously high tide. The morning newspaper of the 18th contained the warnings issued the night before, and all editions of the afternoon paper contained the full warning of the hurricane, and advices as to what might be expected locally. The Peninsular Telephone Co. "Long Distance" cooperated in distributing the warnings, and "Information" answered many calls, as also did the newspaper offices and the Board of Trade. At the residence of the official in charge Mrs. Bennett sat by the telephone all day notifying a long list of people who would be most affected by the gale-contractors, bay-shore residents, etc.,- and after the news had been fully spread, answered innumerable calls from people who sought further information and were not able to get a call through to the Weather Bureau office. This condition lasted until dawn of the 19th.

At the office it was found necessary to exclude the public by placing desks in front of the open doors, so that observations might be taken and other necessary work done, but information was given out as sppedily as possible verbally and by printed card. One man was constantly busy at the telephone, but was relieved at short intervals by a voluntary assistant. A former messenger of the Bureau, Cyril McGraw, gave valuable services Saturday and again Monday, nearly all day. Boy Scouts carried messages to the telegraph office, distributed cards as printed, and made themselves generally useful. The office was not closed from 7:30 a.m., Saturday until 11:15 p.m. Sunday.

The organized forces of the city-the police, fire, and sanitary departments and the Red Cross- were mobilized to be ready for any emergendy, although it appeared that Tampa itself was not to feel the full effect of the storm. The assurance that the tide would not be dangerously high gave much relief, as the high tide had been the worst feature of the 1921 storm.

Locally in Tampa and along the coast and beaches from Tarpon Springs to Sarasota, the warnings, although not issued as far ahead as might be desired were ample, accurate, and fairly well distributed, and undoubtedly resulted in the saving of property and life. Much appreciation has been expressed in the local papers and by individuals. From Sarasota southward, unfortunately, the storm in its full intensity followed so closely upon the warnings that little preparation could be made.

The barometer had been below normal since the 15th, but showed the regular semi-diurnal swing. After the early morning of the 18th it continued steadily downward, reaching 29.59 inches at noon, and the lowest point, 29.36 inches at 5:30 p.m. After that it rose gradually until 11 p.m.(29.43 inches), flattened out until 2 a.m. of the 19th, and then rose to 29.73 inches at noon. It continued low until the 22d.

The first verifying velocity was reached at 12:21 p.m. of the 18th, 28 miles an hour from the northeast. The last was at 10:18 a.m. of the 19th, the storm lasting nearly 22 hours. The average velocity for the 24 hours ending at 12 noon of the 19th was 32.2 miles per hour. The wind continued northeast until after 9 p.m., then turning to the east. and then to the southeast at about 6 a.m. of the 19th. A maximum wind of 50 miles an hour from the east or east-northeast was recorded at 9:33 p.m. of the 18th, and an extreme of 58 at 9:46 p.m. Another 50-mile velocity, this time from the southeast, was recorded at 1:06 a.m. of the 19th, and an extreme of 66 from the east at 12:55 a.m.

The total rainfall during the 18th and 19th was 4.22 inches, at no time excessive according to the register, but much rain must have been blown over the raingage.

The tide was very low, being driven out of the bay and river by the strong northeast winds, as was the case in 1910. At 3 p.m. of the 18th it was 0.8 feet below mean low water, at 6 p.m., 4.0 feet below, and at 9 p.m., 6 feet below (lowest point). High tide on Sept.19 about 2 p.m. reached 4.5 feet above mean low tide, making the extreme range, 10.5 feet. It was high again on the 20th, being 3.9 feet above mean low. In 1910 the tide went 8 feet below mean low tide, and in 1921 it went 10.5 feet above.

The damage locally has been estimated as around $100,000. This includes the loss of the telephone company, about $10,000, and the electric light company perhaps twice that amount. The remainder of the loss is in small amounts, althou quite widely distributed, mostly windows, awnings, trees, and roofing, and damage to plaster and furniture by rain. No house was demolished, but many will need ne roofs.

The loss at St. Petersburg will be about $100,000. Bradentown, Palmetto, and Manatee together will have a loss of a quarter million dollars. First estimates for Sarasota gave $1,500,000, later estimates reducing this to $200,000, but this latter figure probably is too low. Punta Gorda suffered heavily, many roofs being blown off and a number of buildings being entirely destroyed. The wind velocity was estimated at 80 miles an hour, and the top of the storm warning tower was blown off, breaking the upper lantern. The next day the tide rose six feet above normal, flooding portions of the city. The lowest barometer was 28.70 inches at 5 p.m. Near Captiva Island, between Punta Gorda and Punta Rassa, a reading of 28 inches was reported.

Ft. Myers suffered most of the west coast cities, and the loss has been estimated at $3,500,000. It is hoped that later reports will discount this. Many buildings were destroyed or greatly injured and two churches were damaged beyond repair. Very few houses were left without some damage and the beautiful tropical trees were left wrecks. High water accompanied the storm, according to newspaper reports, reaching 4 to 6 feet above normal and flooding certain sections of the city. Sunday morning there was a tide of 3 feet above normal. The lowest barometer was 28.14 inches, and the wind reached an estimated velocity of 80 miles an hour. The anemometer wires broke when the wind was 70 miles.*******

Punta Rassa was wrecked. Points south of Ft.Myers have not been heard from, but that section of the coast has few and small settlements. Boca Grande and Tarpon Springs have not yet been heard from. Damage at Boca Grande must have been very heavy.

At Egmont Key, the waters of Tampa Bay were driven over the island, almost as bad as in 1921, but there was little actual damage. At Clearwater much damage is reported. Perhaps $100,000 would cover the damage at Clearwater, Tarpon Springs and Largo, as well as neighboring beaches.

No lives were lost in Tampa during the storm, nor,as far as I can learn, in any of the cities from Punta Gorda northward. Two were reported dead in Ft. Myers, and 10 in that vicinity from schooners and other craft. Two dead also at Punta Rassa.*******

Shipping was generally warned and heeded the warnings. All vessels carrying wireless got out of the way of the storm, and most of the craft in the harbor remained there unharmed. Fishing schooners and the fruit schooners from the south, as well as sponge schooners, must have been lost-how many vessels and lives it is impossible now to state. Boats in the river and bay listed badly when water went out, but were properly rooked after and not injured with the returning tide. High tide stopped well under the danger mark here.

<u>Apalachicola, Fla.</u> -******The storm reached its greatest intensity here on the 19th. Showers occurred early in the day and continuous rain after 6:52 a.m.; total for the day, 3.27 inches. The wind continued from the northeast, averaging 43.4 miles an hour for the day. It reached a maximum velocity of 40 miles an hour between 2 and 3 a.m., increasing to 50 miles at 5:03 a.m., and 60 miles at 12:19 p.m. The highest velocity was 64 miles an hour from the northeast at 1:16 p.m., with an extreme velocity of 68. The wind became easterly by 6 p.m., and southeast by 9 p.m., continuing from that direction until after midnight. Maximum velocities up to 60 miles an hour occurred as late as 5:10 p.m., then from 43 to 52 the remainder of the day. The barometer reached its lowest point, 29.57 inches, about 5 p.m. The tide was down to -0.4 foot at 6:30 a.m.; it rose during the day reaching 3.4 feet at 6:30 p.m., and continued to rise during the night.

On the 20th there was steady rain until about 7 a.m., then showers until mid-afternoon; total for the day, 1.27 inches. The wind was gusty during the day, continuing from the southeast until 7 p.m., then from the south; average velocity for the day, 36 miles an hour. The maximum velocity was 52 from the southeast at 12:26 a.m., with hourly maxima of 40 or over until 4:10 p.m., then decreasing hourly to 28 miles from the south at 9:51 p.m. The barometer rose slowly during the day, reaching 29.70 inches by midnight. The tide was estimated at 4.2 feet at 6:30 a.m.; it was then overflowing low ground along the water fro The highest water was estimated at 4.6 feet at 11 a.m., with highest waves runnir possibly to 5.0 feet, portions of Water Street being then from 6 to 8 inches unde water. The tide receded slowly during the remainder of the day, falling below 4 feet at 6:30 p.m.

****

The advance warnings regarding the approaching storm were given wide distribution. The telegraph operator kindly opened his office at 10 p.m. Saturday night to receive any weather message from the Central Office, hence the order to hoist hurricane warnings was received promptly. The lights were changed to hurricane warning and copies of the warning were given to many stores still open; it was also telephoned or delivered to many citizens. The telegraph operator again put in much overtime on Sunday to handle weather messages.*****

Very little damage occurred. The high tide and swells damaged one or two of the lowest wharves while the wind did minor damage to a few roofs, small windows and overhead wiring and blew down a few trees. It is estimated $1000 will cover the total damage. Heeding the very efficient advance warnings of the approaching storm, all boats were moved up the river and creeks Saturday to protected anchorages and merchants protected their large plate glass windows Saturday night, sustaining no damage. Many commendations regarding the splendid service rendered by the Bureau have been received at this office.

Pensacola, Fla. - This hurricane was the most severe storm that has ever visited this locality since the establishment of the Weather Bureau, but there was no loss of life, and property damage was not large considering the severity of the storm. The evident reason why this hurricane did not cause the destruction here of some of its predecessors, particularly that of 1906, was that this territory was generally well prepared, having been thoroughly and correctly warned considerably in advance, and that the great part of the destructive winds came from the northeast and east which is off land. Santa Rosa Island, a natural breakwater, also played its part in protecting this city and its environs after the shifting of the wind to the southeast and south.

The center of the hurricane passed southwest of Pensacola, moving northwestward and probably within 25 miles of this city. Advices relative to the approaching hurricane were given widespread distribution by every possible means, and this office was thronged during the 18th and 19th by shippers and interested parties seeking detailed information, and the telephone was kept constantly busy with calls of the same nature. The Naval Air Station and shipping in Pensacola began preparing on the 18th and preparations were carried on actively on the 19th. Airplanes, ships and small craft were placed in supposedly safe anchorage, freight cars were withdrawn from docks and water front, and exposed property generally was protected as far as possible.

No loss of life occurred in this storm-warning district so far as can be learned, and comparatively few injuries were sustained. The S.S.Cardonia, two schooners,atug, and an oil barge are fast aground, a local coastwise steamship and a tug were beached, and eleven fishing smacks were considerably damaged. Numerous small craft were sunk.

All docks were much damaged and the Louisville and Nashville Railroad coaling dock and the Warren Fish Company dock were practically destroyed. The railroad bridge across Escambia Bay was partly carried away, and automobile bridges to Fort Barrancas and the Naval Air Station and on the Spanish Trail across Escambia Bay were considerably damaged, interrupting traffic for several days. Electric light, telephone and telegraph lines were prostrated, and the naval and municipal radio stations were temporarily put out of commission. Rigid awnings, roofs and plate glass windows in the business section suffered considerably and there was some damage from rain and also from high water in the lower sections. Many homes were damaged, but the damage was mostly of a superficial nature such as broken windows, interior damage from the rain, and wind damage to roofs, chimneys, garages, etc. There were very few houses wrecked, but there was considerable flooding in the lower southwestern part of the city which is inhabited mostly by the poorer class. Wood block pavements over the city generally had to be relaid. Many trees, including large live oaks, were uprooted. Two of the city's bathing pavilions were practically destroyed. Water mains on the docks broke, draining the reservoirs, but water was available from other sources, and city water was restored within 48 hours.

The Naval Air Station suffered severe damage, particularly to airplanes and to buildings of semi-permanent character along the water front. The damage at Fort Barrancas was mostly superficial. Damage to property of a serious natur did not occur very far to the eastward of Pensacola.

It is estimated that the damage in this county to standing cotton was abo 35%, the pecan crop, about 40%, and corn, about 25%, but the damage to the satsu orange crop was slight. Pecan and satsuma orange trees sustained practically no injury.

The displayman at Bagdad, Fla., reports that there was no loss of life in his vicinity, but that there was a small amount of damage to roofs, several barg went aground and a small dock was materially damaged. He writes: "We want to thank the Bureau for keeping us fully informed, thus minimizing our loss".

The estimated damage in Pensacola and vicinity is as follows:
Shipping..................................$800,000
Other property........................... 200,000
L.& N.R.R. eastward from city limits to and
including Escambia R.R. bridge............ 375,000
Airplanes and boats at Naval Air Station..1,000,000
Buildings and hangars at Naval Air Station.2,000,000

During the forenoon of the 19th the pressure began falling slowly, and a steady fall of about 0.05 inch per hour set in about 10 p.m. of that date and continued until about 2:30 p.m., of the 20th when the pressure began to fluctuat violently, sometimes as much as 0.15 inch in a few seconds, and these fluctuatio continued for about two hours. The lowest pressure, 28.56 inches, was recorded a

3:10 p.m. This reading is within 0.05 inch of the lowest at this station, recorded in 1917. The pressure rose at about the same rate as it had fallen.

The wind direction was steady northeast for many hours preceding the storm and continued northeast until after 6 a.m. of the 20th; east-northeast until after 9 a.m.; east until about 11 a.m.; east-southeast until about 12:30 p.m southeast until about 3:00 p.m ; south-southeast until about 5 p.m.; and south until after midnight when it again became south-southeast.

The wind was increasing steadily during the 19th and reached hurricane force about 5 a.m. of the 20th, and winds of hurricane force continued for about 17 hours. Between 7:30 a.m. and 11:37 a. m. of the 20th a rate of 100 miles or higher was maintained, and at intervals after that time until about 6 p.m. An extreme velocity of 152 miles was recorded at 7:26 a.m. The station anemometer broke off at the oil hole at 1:47 p.m., and previously the special one-sixtieth-mile contact anemometer went out of commission, due to the breaking of the spindle, consequently the velocity had to be estimated after 1:47 p.m. There is no question, however, but that the highest velocity occurred before the record was lost. The average velocity for the 20th, (partly estimated data) is 76.2 miles an hour. Never before have winds of hurricane force been recorded at this office for so long a time, and never before has the wind maintained a velocity of 100 miles per hour for more than an hour, and only once, in 1916, has the maximum velocity of this storm been exceeded and that record was estimated. Winds of hurricane force were reported eastward to St Andrews, Fla.

Tides were but little above normal until after midnight of the 19th-20th, and at 2 a.m. of the 20th the stage was only 1.3 feet above normal, but thereafter the water rose rapidly and reached its highest stage about noon of the 20th. At 7 a.m. of the 20th the tide was 5 feet above normal, at which time the U.S.Coast and Geodetic tide gage ceased recording. The water remained high until after 2 p.m. when it receded considerably. The water rose steadily in the face of northeast winds of hurricane force, indicating that the storm center was preceded by a moderate tidal wave, as the highest water occurred before the winds became true southeast, or off the Gulf. The high stage of 9.4 feet above mean sea level has been accurately determined since the storm. This stage is 0.6 foot below the high water recorded in 1906, but reliable persons who experienced the 1906 storm assert that the water was higher this year than in 1906, probably by two feet. The U. S. Coast and Geodetic tide gage was not in operation, of course, in 1906, and probably an inaccurate base level was used in computing the level of the storm tide of that year. Reports indicate that the tide did not reach as high a level at the Naval Air Station as at Pensacola, while proceeding eastward in Pensacola Bay much higher levels were reported. The Bagdad Land and Lumber Company at Bagdad, Fla., reports a tide of 14 feet. Valparaiso reports a tide of about 4 feet above normal; St. Andrews, 6 feet; and Port St. Joe about 4-1/2 feet. Reports from the Gulf Beach, about 20 miles southwest of Pensacola indicate that no high water was experienced there.

BAROMETER READINGS ETC., MADE AT

PENSACOLA, FLA.

ept. 19, 1926.			
7:00 p.m.	29.61	NE	50
8:00	29.60	NE	51
9:00	29.60	NE	52
10:00	29.58	NE	52
15	29.56	NE	50
30	29.55	NE	56
45	29.54	NE	56
11:00	29.51	NE	58
15	29.50	NE	58
30	29.48	NE	55
45	29.48	NE	59
ept. 20, 1926.			
12:00 mid.	29.47	NE	59
15 a.m.	29.45	NE	64
30	29.45	NE	62
45	29.46	NE	59
1:00	29.45	NE	57
15	29.43	NE	64
30	29.39	NE	55
45	29.38	NE	60
2:00	29.34	NE	63
15	29.35	NE	66
30	29.35	NE	60
45	29.33	NE	62
3:00	29.33	NE	68
15	29.30	NE	64
30	29.29	NE	70
45	29.28	NE	68
4:00	29.28	NE	65
15	29.26	NE	66
30	29.24	NE	66
45	29.25	NE	70
5:00	29.22	ENE	76
15	29.21	NE	72
30	29.18	NE	74
45	29.16	NE	80
6:00	29.16	NE	83
15	29.13	NE	90
30	29.13	ENE	90
45	29.10	ENE	96
7:00	29.10	NE	96
15	29.10	ENE	90
30	29.09	ENE	99
45	29.07	ENE	100

Sept. 20, 1926.			
8:00 a.m.	29.05	ENE	104
15	29.03	ENE	104
30	29.02	ENE	103
45	29.01	ENE	104
9:00	29.00	ENE	104
15	28.99	ENE	103
30	28.96	E	108
45	28.94	E	106
10:00	28.90	E	109
15	28.90	E	112
30	28.88	E	108
45	28.86	E	108
11:00	28.85	ESE	112
15	28.83	ESE	108
30	28.82	ESE	113
45	28.80	ESE	116
12:00 noon	28.79	ESE	62
15 p.m.	28.79	ESE	58
30	28.76	SE	62
45	28.76	SE	76
1:00	28.76	SE	76
15	28.74	SE	80
30	28.70	SE	82
45	28.68	SE	81
2:00	28.68	SE	70*
15	28.70	SE	70*
30	28.64	SE	100*
45	28.68	SE	100*
3:00	28.68	SSE	110*
15	28.70	SSE	90*
30	28.72	SSE	90*
45	28.71	SSE	70*
4:00	28.76	SSE	80*
30	28.79	SSE	100*
5:00	28.83	SSE	100*
30	28.88	S	90*
6:00	28.89	S	100*
30	28.92	S	70*
7:00	28.96	S	80*
8:00	29.06	S	55*
9:00	29.13	S	60*

* Estimated, after anenometer was destroyed.

Mobile, Alabama.- The hurricane of September 20-21 was characterized by two barometric minima lower than any previously recorded pressure at Mobile; by a longer duration of destructive winds than any former storm on record, although there are three others with higher wind velocities; and by an unprecedentedly low tide, which was followed by high water.

The first advisory warning on September 14, announcing the tropical disturbance some distance northeast of St. Kitts, was given the usual publicity; subsequent warnings were given wider distribution, especially through the Mobile Register and the Mobile News Item. The newspapers used prominent headlines when the storm entered the Gulf; and a special noon edition of the Mobile Register was issued on September 19 in order to give further warning. Timely advance warnings were mailed to a few places on the seashore that are not easily accessible in inclement weather.

The first hurricane warning was received at 9:30 p.m., September 19, and like the order to continue the warnings, issued on the succeeding day, was given extraordinary dissemination. In spreading the information by telegraph and telephone, the office had the cooperation of the Tropical Radio Telegraph Company, the assistant superintendent of the Louisville and Nashville Railway, the manager of the Southern Bell Telegraph and Telephone Company, the superintendent of terminals of the Mobile and Ohio Railway, and the manager of the Home Telephone Company of Mobile; and other parties aided in carrying the warning by boat and automobile. The service was done without expense to the United States; and the distribution of the warning was so effective that no loss of property occurred that could have been averted.

The work of the Bureau has been the subject of laudatory articles by the press, and civic organizations have passed resolutions of commendation.

Meteorological Conditions.

Radiating cirrus clouds moving 7.3 meters per second from the south were observed at 12 noon, September 18. The apex of the angle indicated by their convergence was 345 degrees azimuth; and they were visible until about 4 p.m., when the point of the angle was about 348 degrees azimuth. At nightfall there were nine tenths cirrus clouds moving from the east. On September 19, from the beginning of the day until about 10 a.m., nearly the entire sky was covered with interwoven cirrus clouds moving from the southeast, the velocity at 7 a.m. being 10.4 meters per second. No clouds of the cirri group were visible afterwards, and the sky became overcast with lower clouds at about 3 p.m. The blue of the sky was very pale at sunrise September 19; and there was an almost total absence of the reddish tints, ascribed to the decomposition of light and said frequently to precede storms, at the hours of sunrise and sunset of the days prior to the storm.

The hourly wind movement was 75 miles or more from 9 a.m. to 8 p.m., September 20; a maximum velocity of 94 miles an hour, from the north, occurred during the five minutes ending at 12:15 p.m.; and ten minutes later, the extreme velocity, or the mile made in the least time, was at the rate of 100 miles an hour. The first maximum velocity of over 25 miles an hour occurred at 8:36 a.m., September 19, and the last at 5:40 p.m., September 21; maximum velocities exceeding 40 miles an hour began to occur at 1:05 a.m., September 20, and the last was recorded at 9:35 a.m., September 21; a maximum of 75 miles an hour was first registered at 8:40 a.m., and last occurred at 9:35 p.m., September 20. The wind direction was northeast during the 24 hours ending at 8:30 a.m., September 20, except north-northeast at intervals; then during the period of highest velocities, it was from the north, but with occasional shifts to the northeast prior to 12 noon, and to the north-northwest from 4 p.m. to 8 p.m. It was from the northeast from 9:30 p.m. to 11 p.m., September 20, then from the east until 3:30 a.m., September 21, veering to the southeast by 5:15 a.m.; and it was mostly from the south after 1:15 p.m.

There was a falling tendency of the barometric pressure from September 16, the fall being more marked after 10 a.m., September 19, and it became very rapid after midnight. A minimum of 28.77 inches occurred at 4:30 p.m., September 20, then there was a slight rise until 28.82 was reached at 7:30 p.m., which was followed by a fall to the lowest minimum, 28.76 inches at 9:30 p.m., after which there was a rapid rise.

The rain became heavy about 11 p.m., September 19, and the fall was very heavy at times from 9 a.m., September 20 until after midnight. The total amount of precipitation during the storm was 9.94 inches; of this amount 7.61 occurred on the 20th, and 2.30 inches on the 21st.

The Tide in The Mobile River.

There was a steadily decreasing tide with the northerly winds until an unprecedented low stage occurred at 2 p.m., September 21. The water was 0.5 foot above mean low tide on the harbormaster's gage at 11:30 p.m., September 19, Measurements by the observer of the depth of the water at points of the river bottom reported by different parties as having been above water level indicate a minimum stage of 9.7 feet below mean low tide. This unusual condition became troublesome and caused slight damage to boats that had sought shelter at Twelve Mile Island, upriver from Mobile, as it increased the height of the river bank above water, and the swaying of the trees caused large sections of ground with timber to slide into the river. The tide was reported to be rising at 1 a.m., September 21, and a maximum stage of 5.3 feet was reached at 11 a.m. It fluctuated slightly and remained within a foot of the highest stage until about nightfall.

Damage by The Storm.

No substantial buildings were demolished, but chimneys, sheds, fences, sign boards, and poles of the electric systems were blown down, glass windows were broken in and roof coverings torn off, admitting the heavy rain which injured the interior of houses and stocks of goods. While the individual damages were generally not great, few houses escaped some injury, and the monetary losses from the widespread damage by wind and rain will probably aggregate $350,000. The exceedingly low tides resulted in the retaining piling of some of the older docks giving way, and the damage is estimated at $60,000. There were 35 steamships, 18 sailing vessels, 4 steamboats, and numerous tugs, barges and smaller boats in port. The total damage to these is reported as a few hundred dollars, which indicates the effectiveness of the protective measures taken.

In Mobile and Baldwin counties outside of Mobile equally destructive effects to property from wind and rain occurred as in the city, but the greatest monetary losses were due to damage to crops. A small per cent of the pecan trees were blown down and four-fifths of the pecans were blown off. While some of the nuts may be salvaged, they will be of inferior quality, and the monetary loss is estimated at

$175,000. The cotton destroyed in the field was valued at $60,000. About one-quarter of the corn crop was lost, and the damage to other crops was also extensive. Oak trees, and pines boxed for turpentine suffered severely. There are some tracts of land with as much as half of the timber felled. An approximate estimate of all losses in the two counties is $1,500,000. The damage by the storm extends a considerable distance inland; and there are reports of the destructive effects of the winds in many localities in southern Alabama. At Jackson, the wind capsized a skiff and three negroes were drowned. These are the only deaths reported as having been caused by the storm.

The Storm on the Mississippi Coast.

The severity of the storm decreased rapidly west of Mobile. Based principally on the reports of the storm-warning displaymen and information received from Mr. Frank Craigie, Gulfport, Miss., the following account of the storm on the Mississip coast has been prepared:

Pascagoula.- The wind was mostly from the north-northeast until about 5:25 a.m., September 21, and then it veered to the south-southeast. Fences were blown down, the wire systems injured and other minor damage done. The total loss within the city limits is estimated at $8,000. The greatest loss in the vicinity was sustained by the crops, especially pecans, one-half of which were blown from the trees.

Biloxi.- The wind was from the northeast and north on September 20. The highest velocity which occurred at about 1:30 p.m., was estimated at 50 to 60 miles an hour. During the early morning of September 21 the wind veered to the east; it was southeast at 10 a.m., and south afterwards. The aggregate of the losses within the city was estimated at $3,000. Outside the city, the pecan crop suffered the greatest injury, about 50 per cent of the nuts having been blown from the trees.

Gulfport.- The wind was mostly from the north-northeast until the highest velocities, estimated at from 55 to 65 miles an hour, occurred about 1 p.m., September 20, then the wind decreased and was blowing from 35 to 50 miles an hour when it gradually backed to north-northwest by 9 p.m., On September 21 the velocity again increased and was estimated at from 55 to 60 miles an hour from the north-northwest at 8:30 a.m., when the lowest pressure, 29.08 inches, occurred. At 9:30 a.m., after a lull of 10 minutes, the wind shifted to east-southeast and was about 15 miles an hour. It did not change much until 11:15 a.m., when it began to veer and increase in velocity. It was from the south-southwest by 6:15 p.m., having reached a velocity of about 55 miles from the south-southeast at 2:30 p.m. The tide fell and it was 3.8 feet below mean low tide at 10 a.m., September 20. It rose on September 21 and reached a maximum stage of 6.0 feet at 1:15 p.m. Only minor damages occurred in the city; the value of property destroyed is about $2,000. About 50 per cent of the pecan crop and 15 per cent of the corn crop were lost.

Bay St.Louis.- The wind came from the northeast all day September 20. Beginning at about 9:30 a.m., September 21 there was a calm for an hour, Before the calm the wind was from the northwest and after the storm center passed it blew from the southeast. The highest wind velocity was estimated at 40 miles an hour. The tide was very low from September 17 to 9:15 a.m., September 21, when it began to rise, and it reached about three feet above normal. The total damage done by the storm was estimated at a few hundred dollars.

New Orleans, La.- Advisory messages from the Central Office during the progress of the hurricane before reaching Florida were given the usual wide distribution to shipping by radio telegraph and by direct advices to local shipping interests.

****

Advices issued from Washington during the day on September 18 stated that the hurricane was continuing its movement across Florida into the Gulf, and these advices were given very wide publicity to reach all interests centered in New Orleans, with special attention to shipping, fishing, and coast engineering concerns. Beginning on this day, vessels to sail over southeastern routes out of New Orleans were advised to remain in port.

When the hurricane warning covering the Louisiana coast from the mouth of the Mississippi River eastward was received at 9:35 p.m., on the 18th, all plans were already laid to secure the widest possible distribution in the threatened area. Most of the exposed coastal section has telephone communication, and the hurricane warning was sent to every important point of information throughout the river area below New Orleans, the marshes of St. Bernard and Plaquemines Parishes, and the low-lying sections about Lake Pontchartrain. The superintendent of the New Orleans and Mobile division of the Louisville and Nashville railroad, which has very large interests in the coastal area east of this city, was reached in person, and he placed all the agencies under his direction in motion, not only to warn exposed localities along his line, but also to have every preparation made to meet the storm situation as far as possible as it developed. The officials of the Illinois Central railroad, which is also directly and greatly concerned with adverse storm conditions along the shore of Lake Pontchartrain northward from New Orleans, were likewise advised. Ferry services near the city were warned. The Public Service and telephone companies were advised of threatened danger to their lines. Contractors and dredging interests having large values in floating equipment on Lake Pontchartrain and in Mississippi coastal waters were reached on Saturday night or Sunday morning and warned to take every precaution with equipment, much of which is slow moving and requires considerable time for its protection.

The following message reporting action taken in distribution of the hurricane warning was sent by telegraph to the Central Office at 11:20 p.m., September 18th:

> "Hurricane warning distributed all telephone exchanges Mississippi Delta below New Orleans and points on Lake Pontchartrain. Warnings being given widest possible distribution radio and wire services cooperating. Sent to Fishers on Bayou Barataria to be distributed to Grand Isle and other points that region as advisory". (Signed) Cline.

As a result of these warnings the interests most concerned took prompt action to move all equipment to safe places or take other possible precautionary measures, and to keep in touch with our office as further developments arose. Special representatives were kept in close communication with the Weather Bureau office. The Louisville and Nashville railroad and long distance telephone officials cooperated most effectively, not only in distribution of warnings, but also by reporting conditions to us as they secured information from the coast to eastward.

The radio services at New Orleans, comprising the radio-telegraph station of the Navy (NAT), of the Tropical Radio Company, (WNU and WBW), and the radiophone station of the Saenger-Maison Blanche organizations, (WSMB), were of the utmost service throughout the whole period of the hurricane. The broadcast service by NAT and WNU in dissemination of advisory warnings before the hurricane reached the dangerous waters about Florida no doubt prevented ships from sailing in courses that almost certainly would have brought disaster to some.

We cannot refrain from mentioning here the especially valuable cooperation rendered by Mr. Frank Craigie of the Craigie Press Syndicate at Gulfport, Miss., who, at his own expense telegraphed observations of pressure, wind, and tide at frequent intervals during the advance and passage of the storm. These reports, filling as they did the gap between New Orleans and Mobile, were of material value in definitely locating the storm center as it approached the coast.

Beginning Sunday noon, September 19th, following issue of the hurricane warnings, the radiophone station WSMB began hourly broadcast service on hurricane bulletins. To do this, the practise of the station in observing Sunday as a silent day was waived and the announcer and his staff came on duty and remained at their post until the hurricane was definitely breaking up over the middle Gulf Coast on Tuesday following. This service was of the greatest value in reaching the general public with authentic storm advices and information, and this office made every effort to transmit promptly to the broadcasting station not only the exact text of warning advices from Washington but also accurate data from regular and special observations. Authentic information was thus kept before the public to combat the innumerable wild rumors that invariably float through a nervous population when hurricane danger threatens.

The newspapers rendered their customary service. News representatives were constantly on duty at the Weather Bureau office during the entire period of the storm after hurricane warnings were received. These men were more than ordinarily careful in their handling of matter for their papers, many times consulting us in the actual wording of their statements, and displaying an earnest desire to be accurate rather than sensational.

As the hurricane moved into the Gulf, apparently on a course toward the mouth of the Mississippi River, we printed a small hurricane bulletin on Sunday, September 19th, which was useful in keeping down the congestion of the public about the doors of the Weather Bureau office, which, in the absence of the usual week-day editions of the afternoon papers, threatened to cause serious interference with our handling of the multitude of duties devolving upon us."

The storm having moved across the Florida Peninsula, the length of fetch over the Gulf needed to develop swells that would produce tides preceding the storm, was short. The rise in the tide on the Mississippi coast and at Burrwood up to Sunday morning, the 19th, indicated that the center of the storm was moving toward the mouth of the Mississippi River. However, on Sunday afternoon, September 19, reports from along the Mississippi coast and at Burrwood showed the tide falling at all points. From these, and attendant weather conditions, we concluded that the whole coast eastward to Mobile was in that part of the storm to the left of the line of advance of the center. Our judgment on this matter was therefore embodied in the following telegram to the Central Office, which was forwarded shortly after 4 p.m., September 19:

> "Superintendent Bose Louisville and Nashville Railroad reports tide falling since noon on Mississippi coast, which, with conditions at Burrwood looks like storm center now advancing toward point east of Mississippi coast." (Signed) Cline.

At 4:30 p.m.(local time) September 19 the order to continue hurricane warnings to the mouth of the Mississippi River was received and immediately given the same full distribution by long distance telephone, radio services, and railroad telegraph, that had been accorded the previous order.

On Monday morning, September 20, as the regular a.m. reports were charted, it was conclusive that the storm center was moving as indicated in the last previous advice from the Central Office and that the Mobile-Pensacola section was the area into which the storm center was advancing. The public schools of New Orleans were due to open their fall term, large business interests in this vicinity were operating under restraint and considering suspension of business, and about 25 ships were waiting at anchor in the harbor at New Orleans or lying at the mouth of the Mississippi River, the expense of delay from this last item alone amounting to more than $20,000 per day. It was evident that New Orleans would be in the left half of the cyclone; this would give northwest to west winds which being offshore would not be dangerous on this coast. Therefore, in order to minimize the losses from further delays, to permit ships to proceed, business to continue orderly operation, and to reassure the general public and permit schools to open, we issued the following bulletin based on the a.m. weather map and the last previous Washington advices:

"Storm center evidently passing inland in Mobile section as indicated in warning issued by Weather Bureau last night and New Orleans is on west side of path. Conditions will not be serious here, but we will have increasing northerly winds to-day. Ships sailing out of New Orleans southward may proceed."

****

The order to lower hurricane warnings on the Louisiana coast at 12 noon, September 20, was received and action taken accordingly.

As the storm moved slowly over the coast section, its energy began to be dissipated and the rainfall area spread to the left and rear of the center, indicating a diminishing cyclone. Dense cloudiness set in at New Orleans about noon Monday, and late Monday night rain began that continued almost without intermission until midnight Tuesday.

****

Storm winds did not occur at New Orleans, except in occasional gusts of less than one mile of movement. The maximum velocity was 27 miles an hour from the northwest on the 21st. The lowest pressure was 29.47 inches at 2 p.m., September 21. Damage of about $50,000 resulted from wave action on Lake Pontchartrain. The New Orleans Public Service suffered small damage from broken wires. Damage to floating property, which might have been heavy on Lake Pontchartrain, Mississippi Sound, and at the mouth of the Mississippi River, was so far as known, entirely prevented by timely precautions taken as a result of the warnings issued. The maximum wind velocity at Burrwood was 48 miles an hour from the northwest about 12 noon, September 20.

Spotted something that isn't quite right?

Let us know and we'll fix it for future reprints. And as a thank you for your diligence, we'll send you a free Sothis book of your choice.

publisher@sothis.press

This page intentionally left blank

Except of course for that line above. And this one. And whilst we have your attention, perhaps you'd like to follow us on Instagram @sothis.press. Go on.

Made in United States
Orlando, FL
04 November 2026

90201485R00018